LIFE ON A FISHING BOAT

a sketchbook

Huck Scarry

Prentice-Hall, Inc.

Englewood Cliffs, New Jersey

First published 1983 by Librairie Ernest Flammarion, Paris.
Original title *A bord d'un Bateau de peche* copyright © Flammarion, 1983.
English text copyright © 1983 by Huck Scarry.
First American edition published 1983 by Prentice-Hall, Inc.

10 9 8 7 6 5 4 3 2 1

Library of Congress Cataloging in Publication Data
Scarry, Huck.
Life on a fishing boat.
Translation of: A bord d'un bateau de peche.
Bibliography: p.
Summary: Records the author's journeys on fishing boats in waters of the world from the North Sea to the North Atlantic and describes the various kinds of fishing boats and the daily lives of the people who live on them.
1. Fisheries—Juvenile literature. 2. Fishermen—Juvenile literature. 3. Fishing boats—Juvenile literature.
[1. Fisheries. 2. Fishermen. 3. Fishing boats. 4. Boats and boating] I. Title.
SH331.15.S3313   1983   639'.22'091631   83-9631
ISBN 0-13-535856-6

Preface

If you have ever been to a fish market, perhaps you have been surprised by the variety of different fish which have been brought back from the sea. There are fish of every size, shape, and color, and many of them are very beautiful.

Sometimes while eating a tasty fish at home I have wondered, "Where does each fish come from, and how are they caught?" I have never caught a fish myself, but I decided that the best way to answer my curiosity was to actually go to sea and find out!

This book is the travel diary of actual fishing trips I made with fishermen on the Atlantic Ocean, off the Brittany coast, and on the North Sea. I soon discovered that there are as many ways to catch fish as there are ways to serve them!

The information you will find in this book is what *I* caught at sea.

The Atlantic

A pair of otterboards

What fun to see a fishing port!
There is the salty smell of the ocean, a healthy breeze,
and, of course, plenty of fresh fish!

This is the port of Le Croisic, in Brittany, where
I started my trip. On one side of the principal street
are handsome old buildings. On the other side there are
colorful little boats. Fishing boats!...and fishermen,
busy repairing their nets and their boats.

There is no fishing done in the port. For that
you must go to sea.

*It's low tide!
The boats sit high on the muddy sea floor.
Tides are created by the moon's
gravitational pull on our planet's oceans.
Tides are especially strong in Brittany.
Here at Le Croisic, you can
only leave and return to port
when the tide is high.*

*The fisherman's day
is not determined by the clock
but by the tides!*

But it is here that I met a friendly fisherman who agreed to take me to sea. "Be here early in the morning!" he warned me.

Indeed, the fisherman's morning begins early! Mr. Berrou, my skipper, was already on board ship at 5 o'clock. "Why do we leave so early?" I asked him with a yawn. Mr. Berrou explained that it depended on the tides. We had to be off while the tide was high.

In the darkness I saw two shadows moving on deck.
Mr. Berrou introduced them to me. They were his sons, unloading yesterday's catch which would be sold at auction today.
Mr. Berrou helped me aboard his ship, *La Madone,* and then disappeared below deck to fire up the powerful diesel engine.
The lines were cast, and we moved away from the quay…

Like most fishermen, Mr. Berrou is the son of a fisherman. His sons are fishermen, and his grandsons are likely to be fishermen, too!

…to the vast ocean.

All the fishermen take advantage of the high water to either leave or return to port.

A Coastal Trawler

As we forged our way out to sea, the sun came up, and Mr. Berrou's bright red ship was unveiled to me. I learned it was called a trawler, and Mr. Berrou proudly described her to me.

Mr. Berrou's ship is a typical small coastal trawler. Built of wood, with a fiberglass cabin, it measures 12 meters long by 4.20 meters wide (39.6 x 13.9 feet). This 15-ton ship is propelled by a 180-horsepower diesel engine.

A trawler is, in essence, a tugboat. But instead of hauling large ships, it tows the trawl behind it.

Trawlers can only fish on the smooth ocean floor. Fishermen avoid any rocks or wrecks that might tear the trawl to shreds!

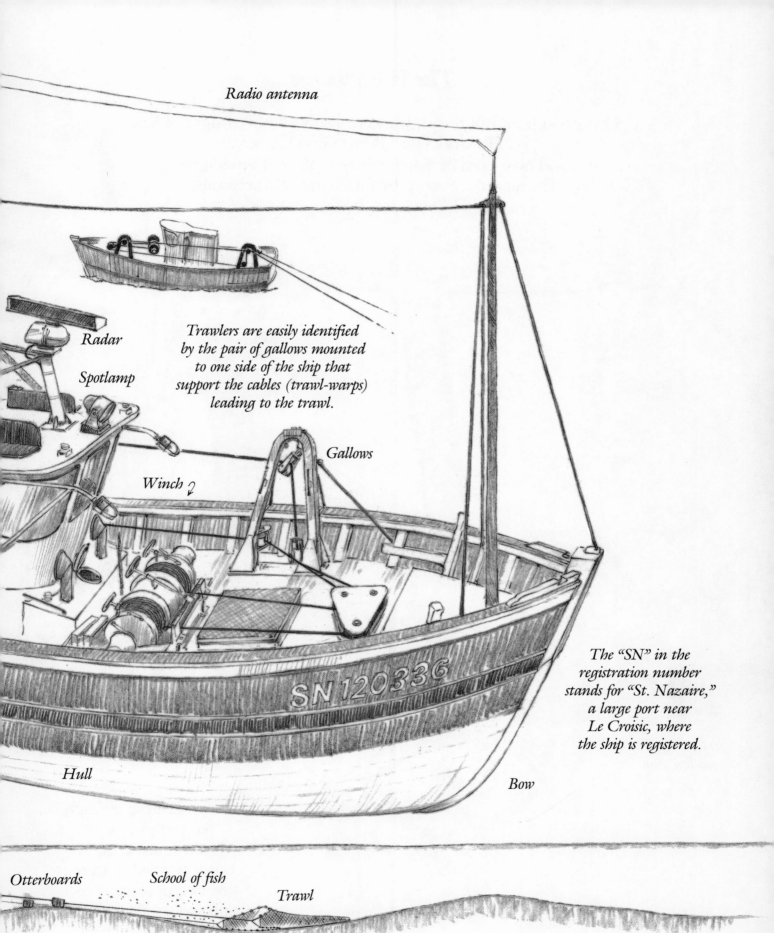

The Wheelhouse

One good hour out to sea, we were also well out of sight of land.
"How does Mr. Berrou find his way,
and how does he know where to fish?" I wondered.
The answers were to be found in the wheelhouse.

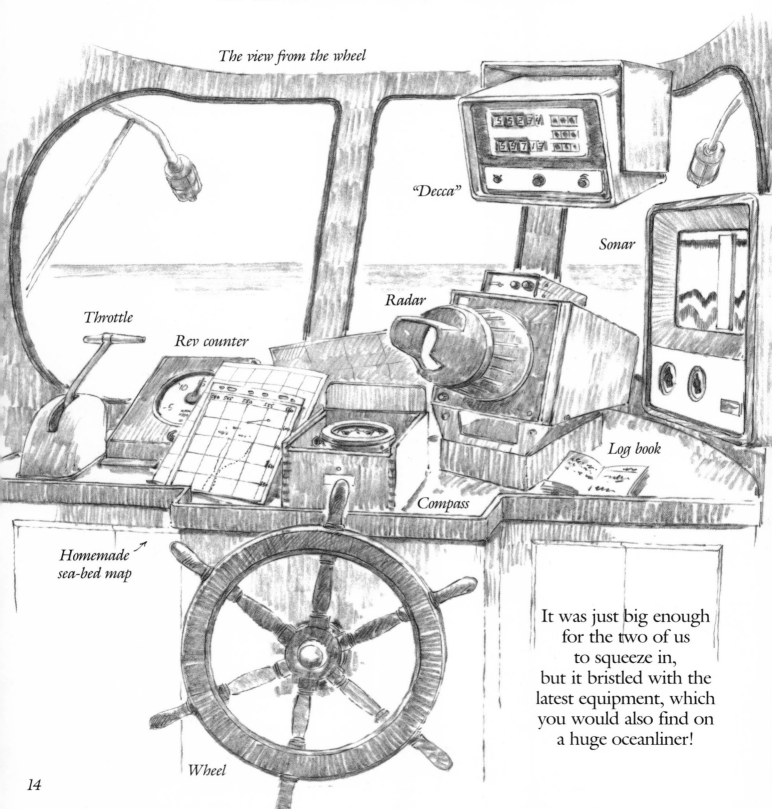

The view from the wheel

"Decca"

Sonar

Throttle

Rev counter

Radar

Log book

Compass

Homemade sea-bed map

Wheel

It was just big enough for the two of us to squeeze in, but it bristled with the latest equipment, which you would also find on a huge oceanliner!

The "Decca" system of navigation is used today around the world. If a pair of radio stations sends out radio waves of the same frequency at exactly the same moment, then a receiver picking up the waves from both at the same instant must lie on a line equidistant from both (line A).

Every fisherman makes little maps of his fishing grounds. The maps are made with the aid of the sonar.

If, however, it takes longer for one transmitter's waves to reach the ship, then the ship must lie on a parabola between the two stations. By arranging two pairs of transmitters that give two sets of intersecting parabolas, it is possible with a Decca receiver to find one's position.

The sonar is one of the fisherman's most important tools. It tells him what is under his ship. Not only does the sonar tell him at what depth the bottom lies, it also shows him the nature of the terrain. The sonar can pick up schools of fish, and a trained eye can even distinguish what kind of fish they are!

Radar, of course, works a bit like sonar but shows instead what is around the ship on the surface. A beam sent out from the radar is reflected back by any objects it encounters, making an image on a special screen.

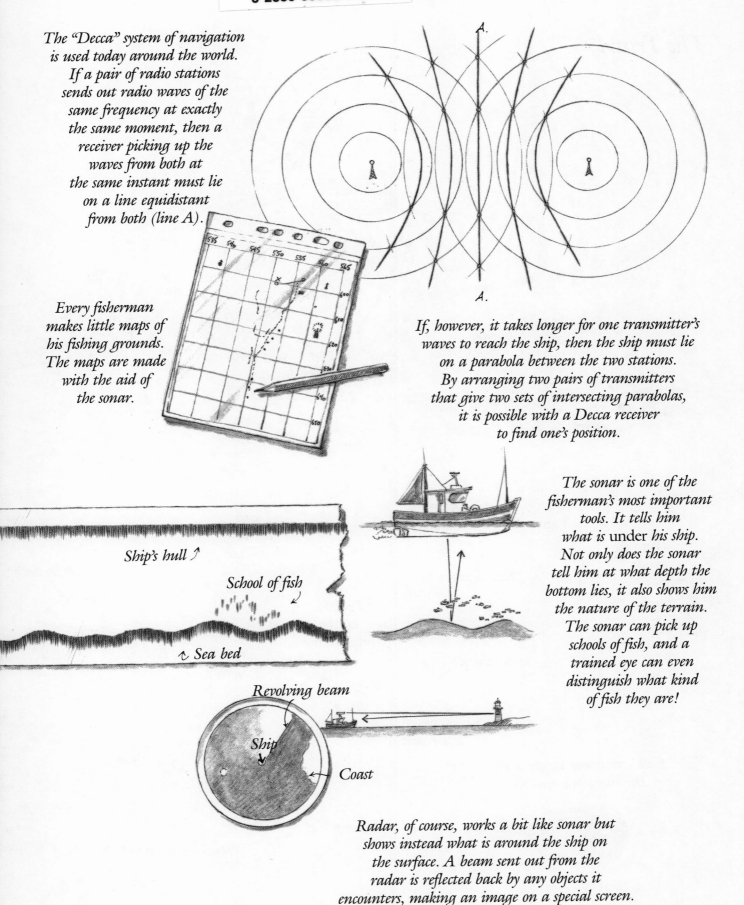

15

The Trawl

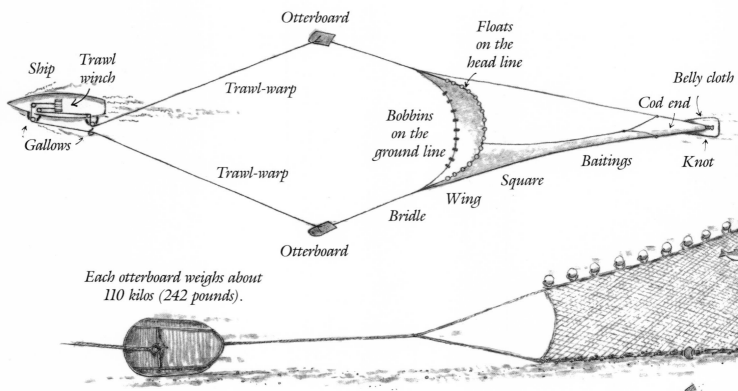

Each otterboard weighs about 110 kilos (242 pounds).

As they are pulled forward, the water's pressure against them pushes them outward, enabling the mouth of the trawl to remain wide open.

Mr. Berrou gave his instruments a good look, checking our position on his map. His hand pulled back on the throttle, and the change in pitch of the engine awakened his sons from their slumber below deck. It was time to drop the trawl…time to start fishing!

The "otterboard trawl" was developed in the last century.

The trawl is the big net that catches the fish. It actually rakes the ocean floor, pulled along slowly by the trawler at the other end of a pair of long sturdy cables. While trawling, I followed our path on the map. Mr. Berrou raked this corner of the ocean as carefully as you would rake autumn leaves from your lawn!

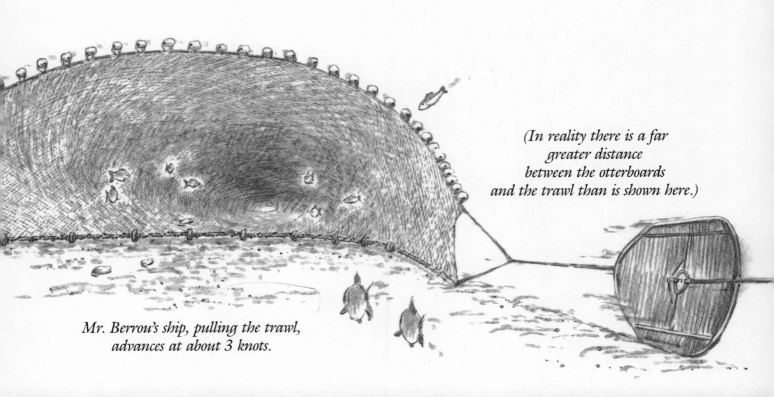

(In reality there is a far greater distance between the otterboards and the trawl than is shown here.)

Mr. Berrou's ship, pulling the trawl, advances at about 3 knots.

After two hours on the ocean floor, Mr. Berrou thought the trawl must be full enough to pull up on deck. A powerful winch is used for this strenuous job.

About 50 kilos (110 pounds) of saleable fish come up with Mr. Berrou's trawl.

Mr. Berrou returns to port every day with 300 to 500 kilos (650 to 1100 pounds) of fish.

I watched the trawl appear near the surface, alongside the hull.
A cord pulled it up, via a pully on the mast, over the deck.
It looked like Santa Claus's bag of gifts!
Mr. Berrou gave a pull to the slipknot at the base of the "bag,"
and the slippery, slithery, slapping, and flipping bounty
of the sea fell onto *La Madone's* deck.

Without a wasted moment the slipknot
was retied, the trawl recast overboard,
and the fishing started again.
On deck, meantime, Mr. Berrou's sons
sorted out the different fish,
placing each kind in a separate box,
like postmen sorting letters.

Then the fish were cleaned
...inside

...and outside.

The Fish

Just what did Mr. Berrou catch? Here is a sample of what he frequently finds in the sea.

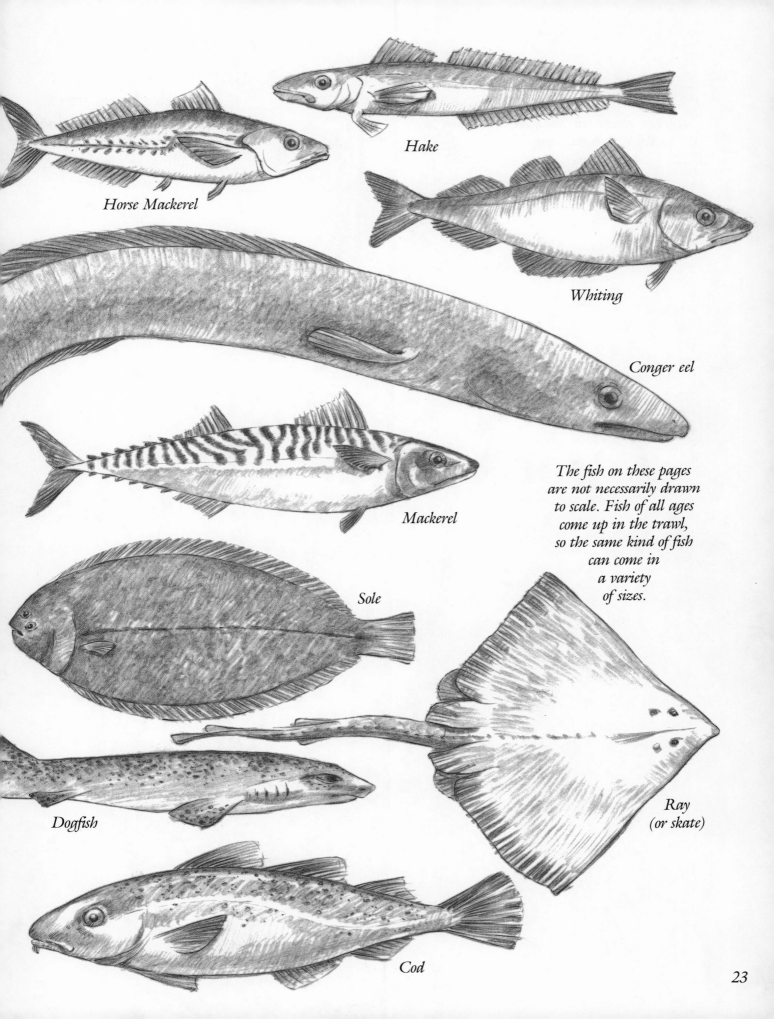

You can almost tell how good a ship's catch has been by counting the number of seagulls in its wake!

Twelve hours after setting out (one full tidal cycle) and several trawling sessions later, it was time to return to port.

The crews from other ships greeted us on the quay with a wave. Everyone lent a hand to bring the catch ashore. Crushed ice would be mixed into the boxes of fish to keep them good and fresh for the auction next morning.

It had been a busy day's work for Mr. Berrou... tomorrow would tell how profitable it had been!

The Auction

The catch of each fishing boat is sold at auction. The auction takes place every morning at 6 o'clock in a special hall.

Fishmongers from the surrounding towns and cities come to buy the previous day's catch.

As the auctioneer presents the lots, the fishmongers place their bids. The top bidder then leaves his card on the box he has purchased to identify it.

Auctioneer

Bookkeeper

Fishmongers

On the top is written "La Mad," which stands for Mr. Berrou's ship.

While the fishermen lay their catch out for inspection, a noisy auctioneer, armed with a baton and a megaphone, sells off the lots (one, two, or more boxes) to the highest bidder.

Here is a sample sales sheet. In the first column are the code names of all the different fishmongers. In the middle column are the lots of fish, and in the last, the prices paid for them.

The sales of fish are recorded in the newspaper.

The lots of fish are weighed.

The fishmongers load their purchases in refrigerated trucks and speed away to market.

A Stern Trawler

A recent development in trawling is the stern trawler, which, as
its name implies, hauls its trawl over the stern instead of
from one side. This particular ship, which belongs to
Mr. Berrou's cousin, specializes in fishing for Norway lobster,
also known as scampi or prawn (*langoustine*, in French).

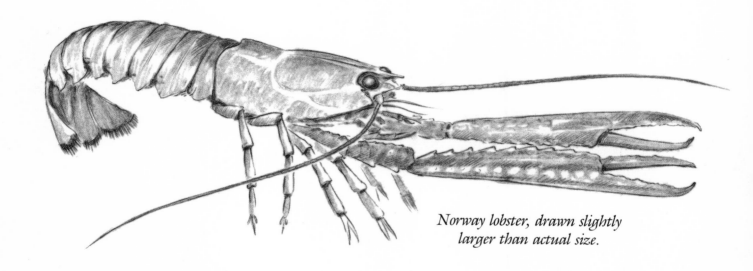

*Norway lobster, drawn slightly
larger than actual size.*

The banks of Norway lobster are only found a good number
of hours out of port, so to make a trip worthwhile,
the skipper goes out for two days.
The catch of the first day is frozen in ice
in the hold, while the second day's catch comes
to port alive in the fish-wells on deck.

What characterizes a stern trawler is the great gallows standing on the stern. The fish-wells which keep the Norway lobster fresh can be seen near the bow.

A Lobster Boat

Crabs and lobsters live on the rocky parts of the ocean floor.
No question of trawling these armored and beclawed animals
amongst the rocks... there would soon be no trawl left!

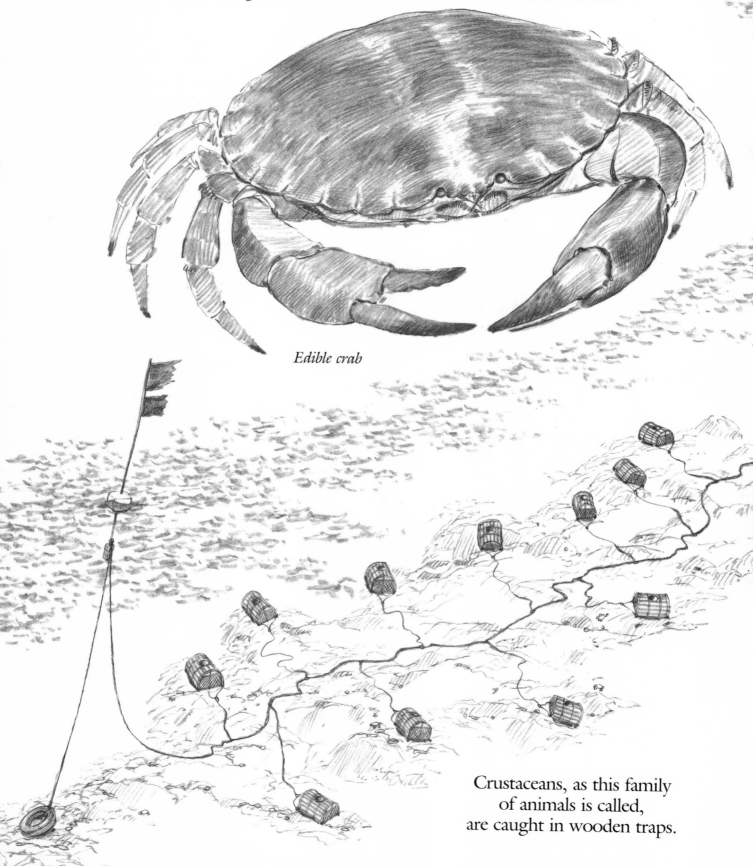

Edible crab

Crustaceans, as this family of animals is called, are caught in wooden traps.

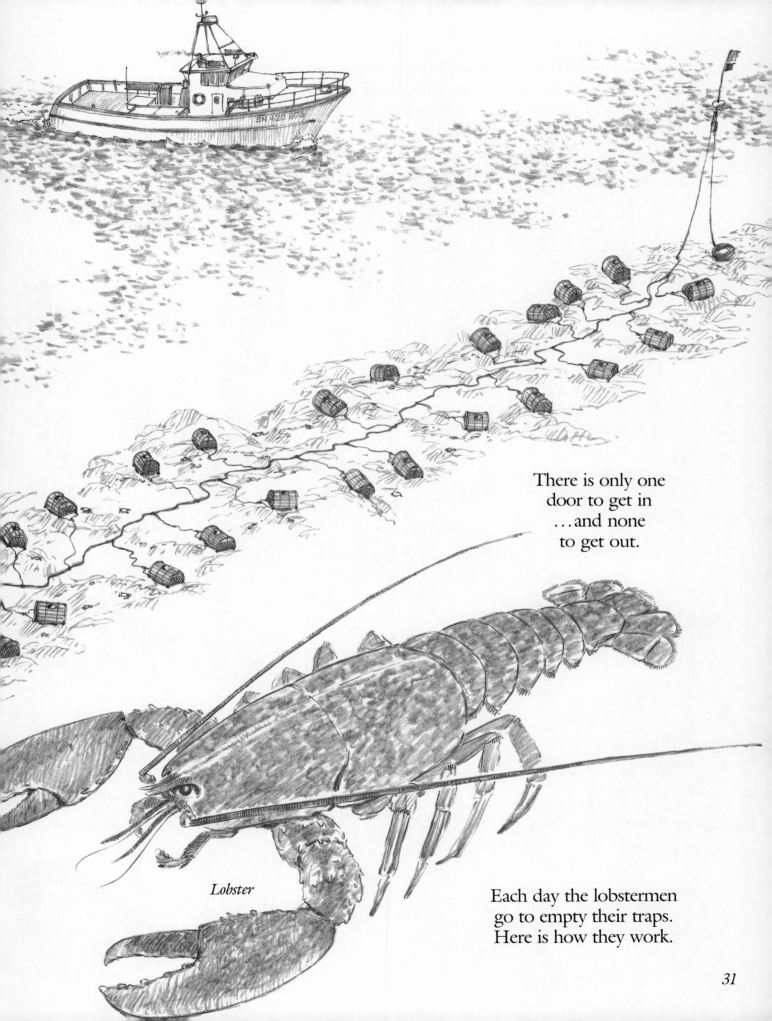

There is only one door to get in ...and none to get out.

Lobster

Each day the lobstermen go to empty their traps. Here is how they work.

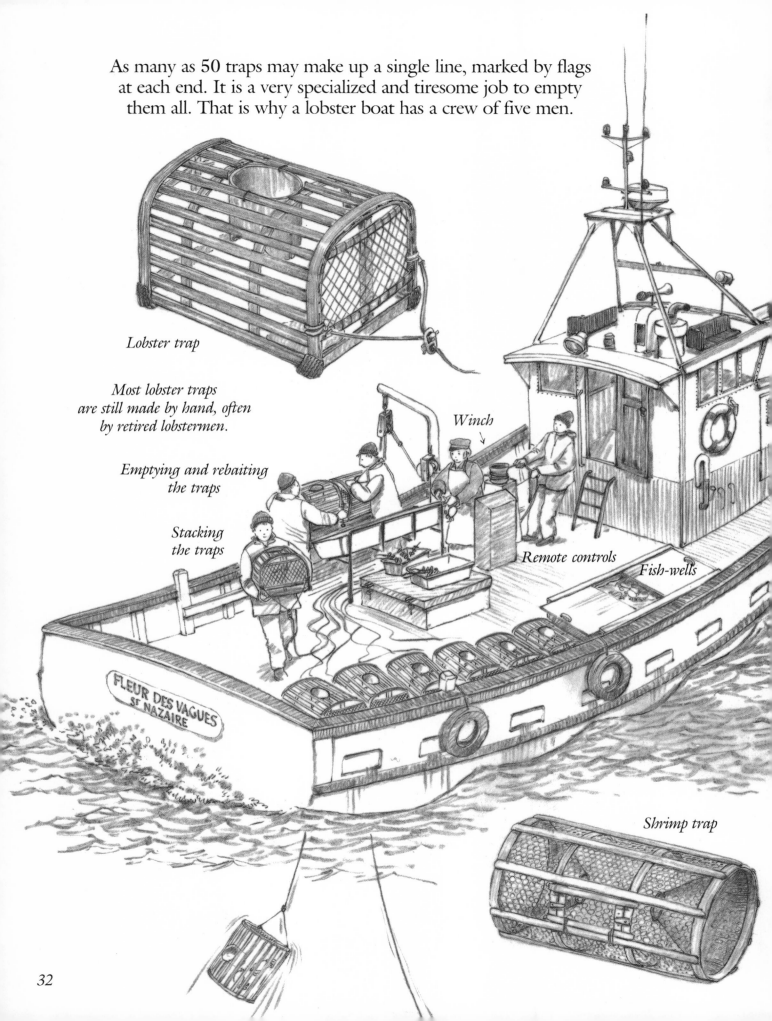

When work is going on, the spacious flat deck of the lobsterboat fills with furious activity. Each crew member mans a particular post. One man with the aid of a winch hauls up the line, a second empties the traps, a third rebaits them, a fourth stacks them, and the fifth cuts the tendons on the crabs' claws, making their pincers harmless.

The traps come up and are emptied at a rate that would have been the envy of Henry Ford himself! No sooner are all the traps on deck than they are cast back into the sea, one after the other. The lobstermen will be back again tomorrow!

*The lobstermen fish for different crustaceans with the different seasons:
June to October is for crab and lobster;
October to February is for shrimp;
and February to June is for spiny lobster,
far out at sea.*

*We emptied 8 lines of 50 traps—
a hard day's work!
When the traps come up, the crabs
are dumped into the fish-wells.*

*Conger eels lured into
the traps by the bait
are notorious housewreckers!*

33

Sardine Boats

Sardines have always held an important place in the human diet.
Pressed and salted, they could be eaten long after they were caught.

The fishing and preparing of sardines used to occupy a large portion of
Brittany's coastal population, especially after Nicolas Appert, in the last century,
invented the technique of *canning* sardines. There were over 100 canning factories
along the Brittany coast alone. And while the men went to sea in
little sardine boats, the women worked ashore, filling little sardine cans!

*Sardines used to be caught from distinctive little boats like this one.
Once on a bank of sardines, a drifting net was cast out behind the boat.
Then the captain would throw out handfuls of roe (fish eggs) to each side of the net.*

*The sardines, in their effort to gobble it up, would get
caught in the mesh. When the skipper thought the net was
sufficiently full, it was hauled in, the fish picked out, and the net thrown
out again. Usually a boat carried a number of nets with
different size meshes, to catch sardines of every size.*

These ladies wear the distinctive embroidered coiffes of Brittany.

In Brittany today
far fewer people are employed
in sardine fishing.
The factories can be counted
on your fingers.
Yet Le Quiberon
is a port where both activities
are still carried out…
and fascinating it is to see.
Sardine fishing is now done at night,
on fairly large ships.
Guy Jacob and his brother
work two of the three sardine
ships found at Le Quiberon.
I joined his crew one night
on the quay, and a
little dory took us to his ship,
Kanedeven, which in Breton
means "Rainbow."

Mr. Jacob's ship measures 15 meters long by 5.20 meters wide (49.5 by 17.2 feet) and is powered by a 230-horsepower engine.

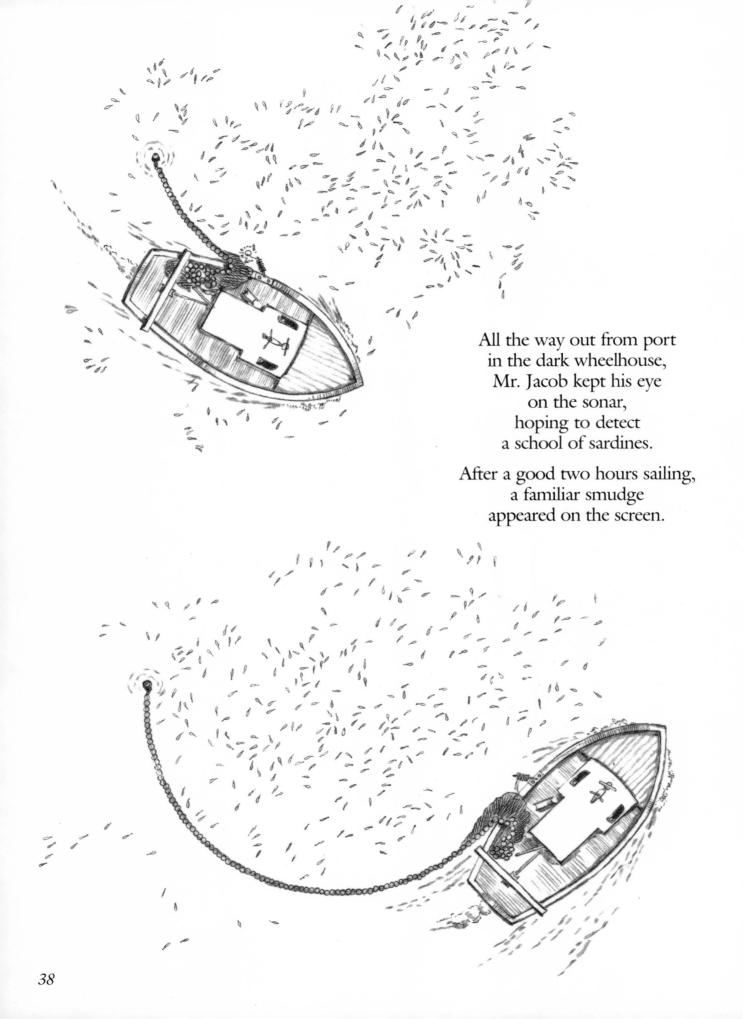

All the way out from port
in the dark wheelhouse,
Mr. Jacob kept his eye
on the sonar,
hoping to detect
a school of sardines.

After a good two hours sailing,
a familiar smudge
appeared on the screen.

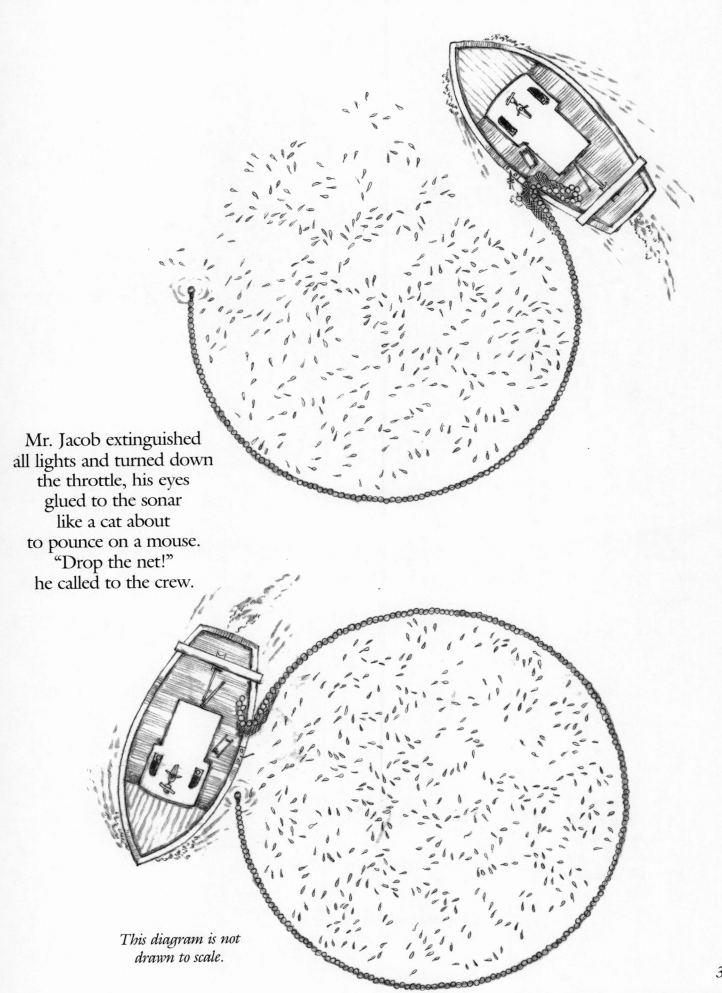

Mr. Jacob extinguished all lights and turned down the throttle, his eyes glued to the sonar like a cat about to pounce on a mouse. "Drop the net!" he called to the crew.

This diagram is not drawn to scale.

39

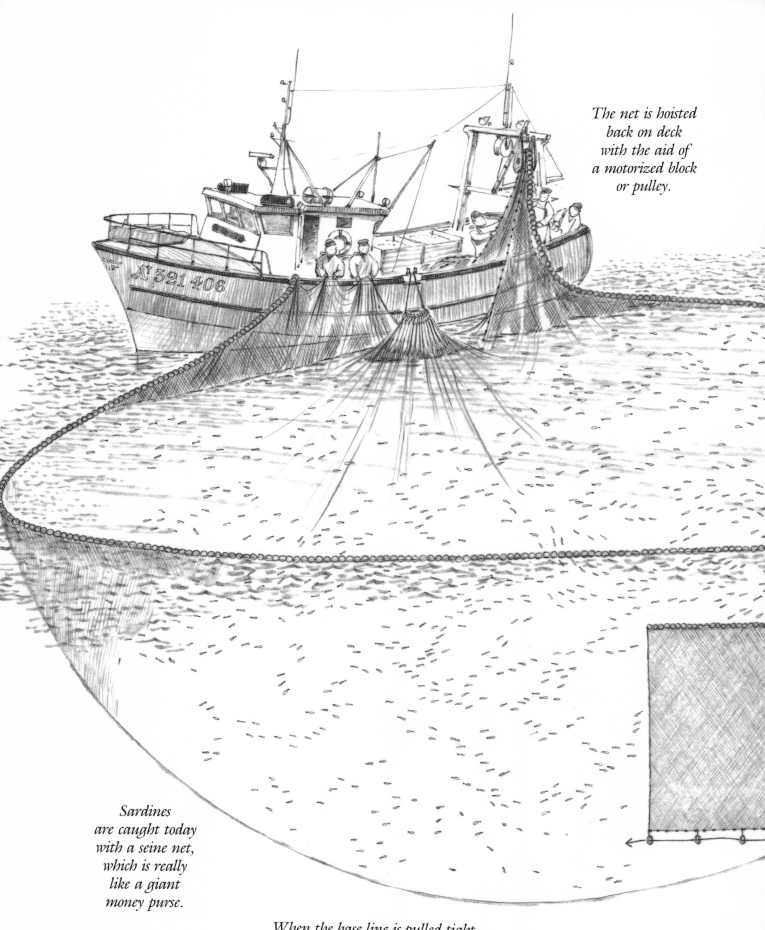

The net is hoisted back on deck with the aid of a motorized block or pulley.

Sardines are caught today with a seine net, which is really like a giant money purse.

When the base line is pulled tight, there is no way for any of the precious contents to fall out!

The sound of ropes, the clank of chains, and the whir of winches cut the darkness. Mr. Jacob carved a large circle with his ship, and the net encircled the school of fish. Before they had a chance to dive—their only escape—the crew hauled tight the base line, closing the big net like a purse.

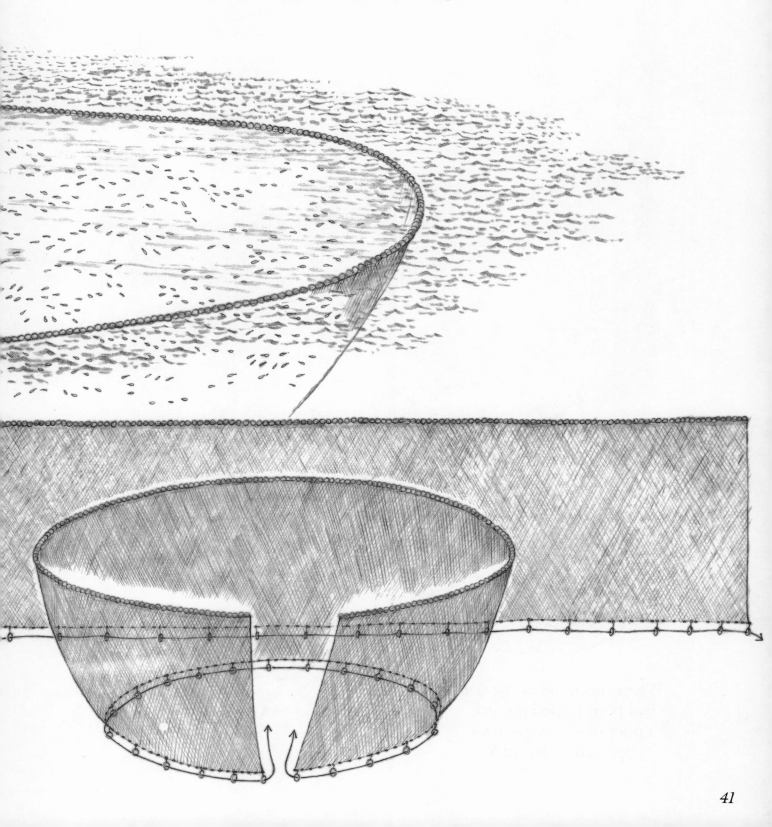

Now that the prey was definitely trapped, Mr. Jacob could light up the deck. At last I could see what was going on. The net was hauled in by the crew with the help of the large powered pulley mounted on a spar.

When almost all of the net had been hauled aboard, a pool of silvery sardines appeared by the hull.

Mr. Jacob manned a giant scoop and delved into the glittering mass, hoisted it over the deck, and emptied the booty onto the ship. Again and again he scooped, until the deck began to resemble a national treasury!

When all was done, he ran back to his sonar to search for the next bank of silver …um, sardines!

In summer Mr. Jacob brings home four to five tons of sardines, and three times that many in winter.

The Canning Factory

When the boxes of sardines arrive at the factory, the fish are removed from the ice and placed in special trays that are dipped in a bath of pickling brine for 16 minutes.

We returned to the port of Le Quiberon early next morning, and I followed the sardines to the canning factory. A lot of work goes into those little tin boxes!

The fish are then washed, dried, and plunged into a bath of peanut oil (which leaves no taste), at a temperature of 120–125°C (248–257°F). The fish then rest for twelve hours to let any excess oil run off.

The fish heads, tails, and fins are snipped off with scissors, and the fish are placed in cans by hand.

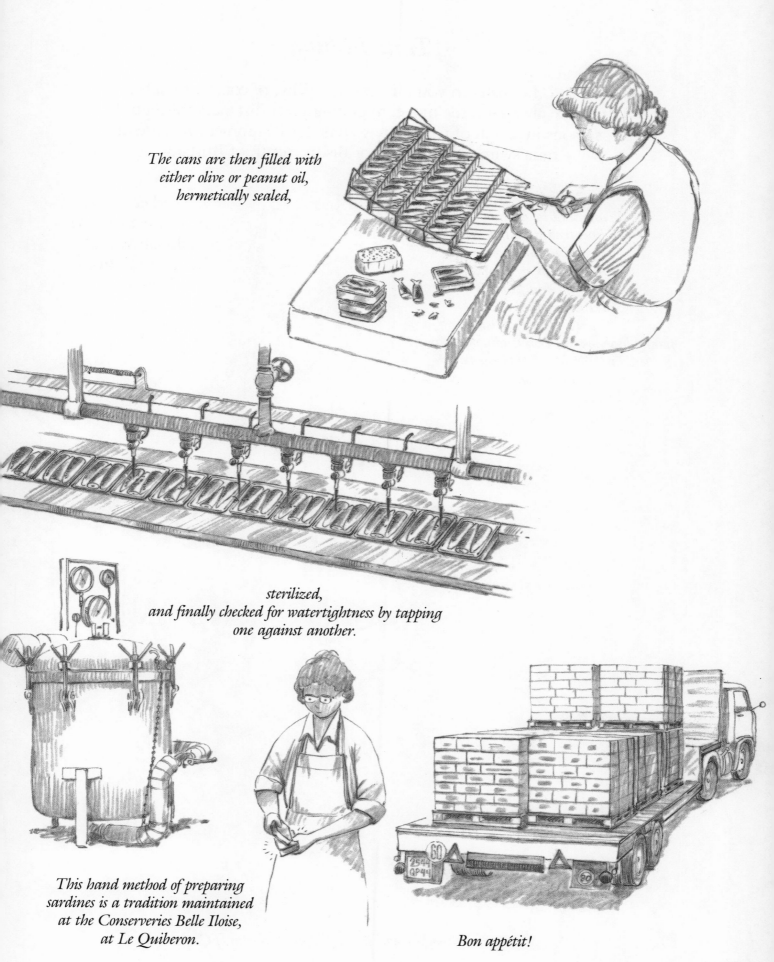

The cans are then filled with either olive or peanut oil, hermetically sealed,

sterilized, and finally checked for watertightness by tapping one against another.

This hand method of preparing sardines is a tradition maintained at the Conserveries Belle Iloise, at Le Quiberon.

Bon appétit!

Tuna Fishing

What other fish can you buy in cans? Why, of course, tuna fish! Tuna are always on the move, migrating great distances through the seasons in search of temperate waters. Each summer one type of tuna swims into the Bay of Biscay, south of Brittany.

Tuna used to be caught on lines trailed behind the advancing tuna boat. Once hauled on board, the powerful tuna was clubbed on the head. Every line on these old fishing boats had its own traditional name, such as "Old Maid," "Goodfellow," and so on.

In earlier times, the distinctive tuna sailboats would then set sail from Brittany.

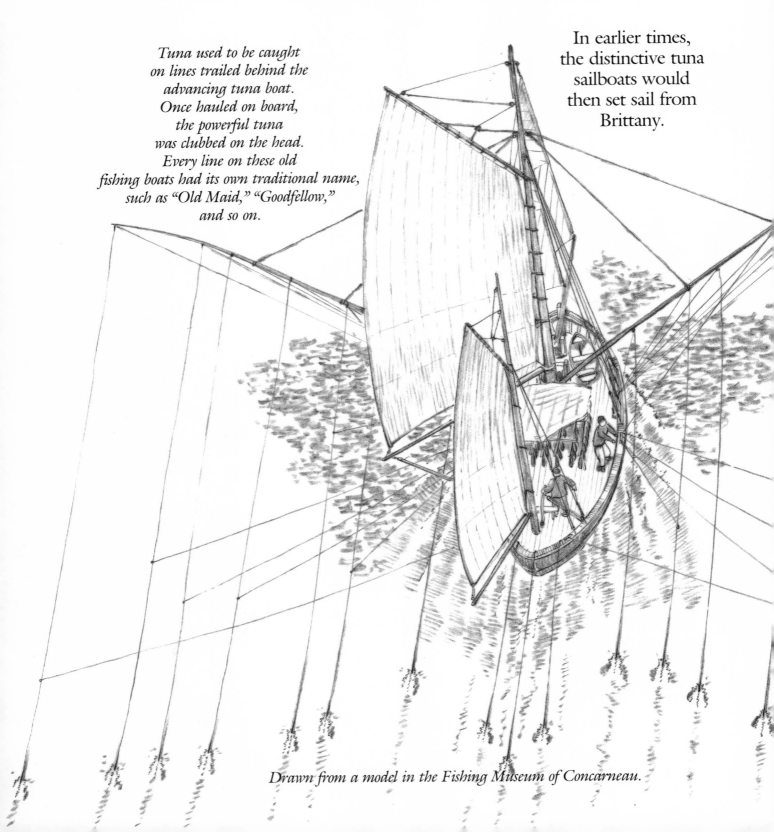

Drawn from a model in the Fishing Museum of Concarneau.

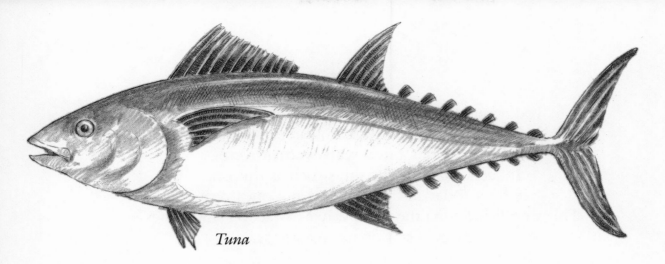
Tuna

Canned tuna

The charming fortified town of Concarneau was home for many tuna boats. It was here in the fine Fishing Museum that I learned how to catch a tuna.

Clipper

The "CC" on the registration number of this ship stands for "Concarneau."

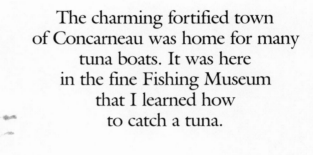

Clippers were first used in the Pacific. Lines of men holding fishing rods stood to one side of the ship, and once on a bank of tuna, all they had to do was put a line in the water for a tuna to bite it... even without bait! It was found that a spray of water on the surface excited the tuna even more and made the catch easier still.

Today most tuna fishing is done in tropical seas, but many ships in African waters still bear the name of Concarneau on their stern. Indeed, magnificent ships such as this one are built today in the shipyard at Concarneau. This ship fishes tuna the same way Mr. Jacob's ship fishes sardines... but on a tuna-size scale!

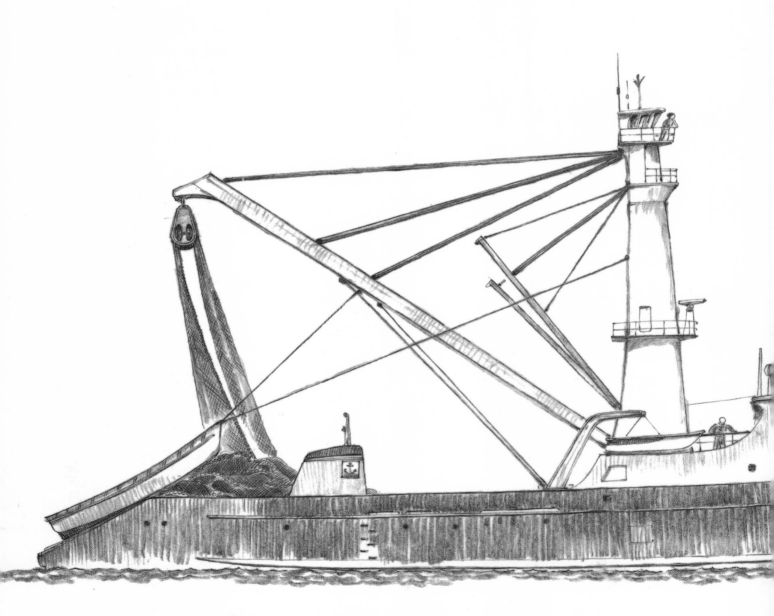

Drawn from material exhibited in the Fishing Museum at Concarneau.

The auxiliary boat that helps place the seine net correctly around the bank

This is how tuna is caught today. The helicopter goes out to spot banks of tuna from the air. A crow's nest atop the great mast is also used for spotting.

Mr. Berrou's ship is shown here alongside for comparison. Sometimes as much as 150 tons of tuna are brought on board from one net! The tuna are immediately dumped in freezers in the hold of the ship.

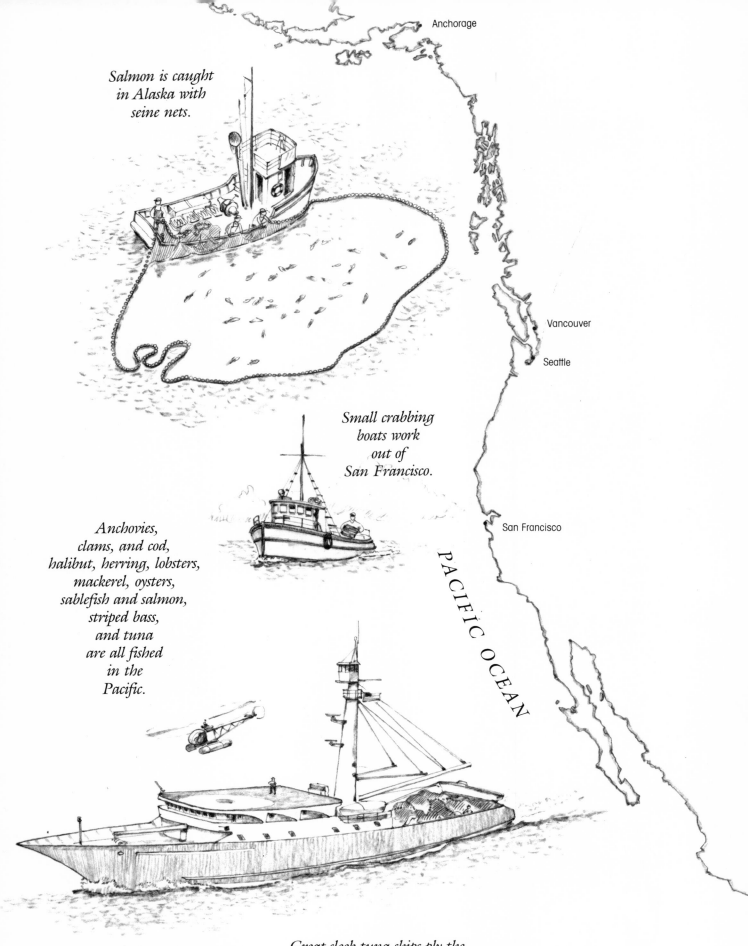

Salmon is caught in Alaska with seine nets.

Small crabbing boats work out of San Francisco.

Anchovies, clams, and cod, halibut, herring, lobsters, mackerel, oysters, sablefish and salmon, striped bass, and tuna are all fished in the Pacific.

Great sleek tuna ships ply the warmer waters of the Pacific.

North America

Fishing, of course, is also an important activity along the long American seacoasts.

Anchovies, cod, crabs, haddock, hake, halibut, and herring, perch, scallops, lobsters, and oysters are fished in the North Atlantic.

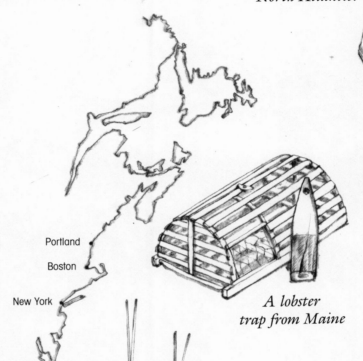

Scallops are dredged off the North Atlantic floor.

A lobster trap from Maine

Portland
Boston
New York
Norfolk

Grouper, menhaden, red snapper, and shrimp are fished in the Gulf of Mexico.

Oysters are picked off the shallow sea bed with long tongs from small dinghies.

Tampa

GULF OF MEXICO

ATLANTIC OCEAN

Trawling for shrimp is a major activity in the Gulf of Mexico and along the southern seaboard.

The North Sea

The waves of the North Sea lap several European nations.
It truly is northern Europe's fishpond!
Here ships from different lands fish the same banks.
Two brother fish living on Dogger Bank and caught by
different boats might wind up on tables in London and in Oslo!

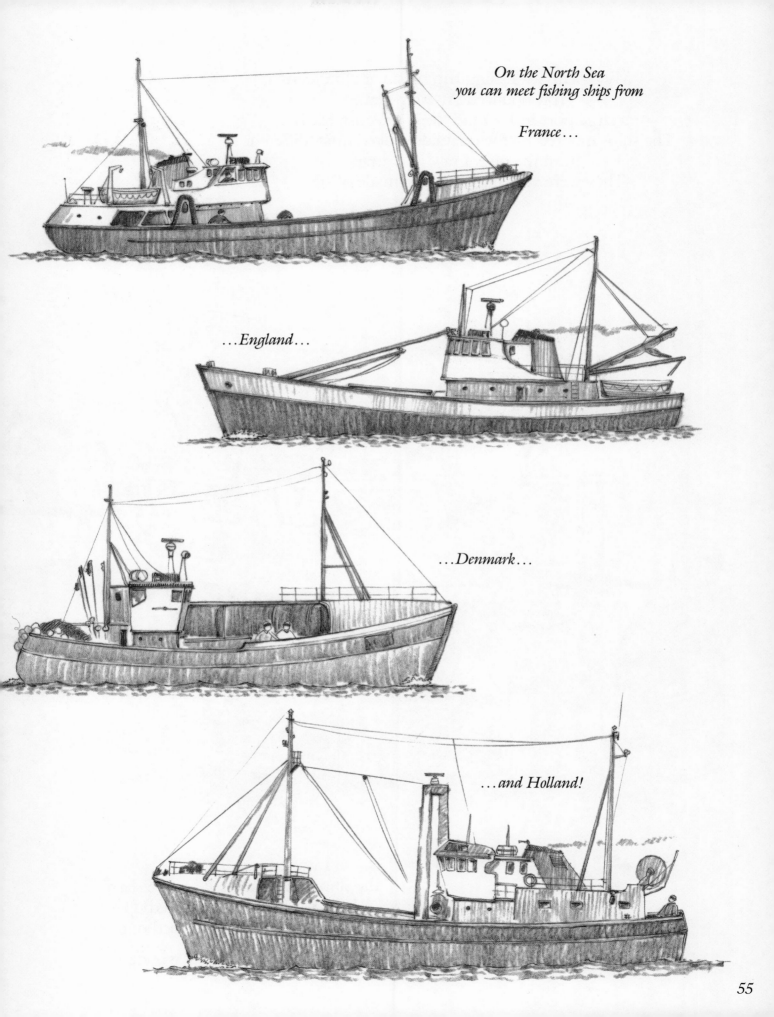

My North Sea fishing trip began at IJmuiden
(pronounced Ay-moy-den),
a large port in Holland, west of Amsterdam.
The ships moored by the quayside looked quite different
from the ones I saw in Brittany.
They were very big and were made of steel.

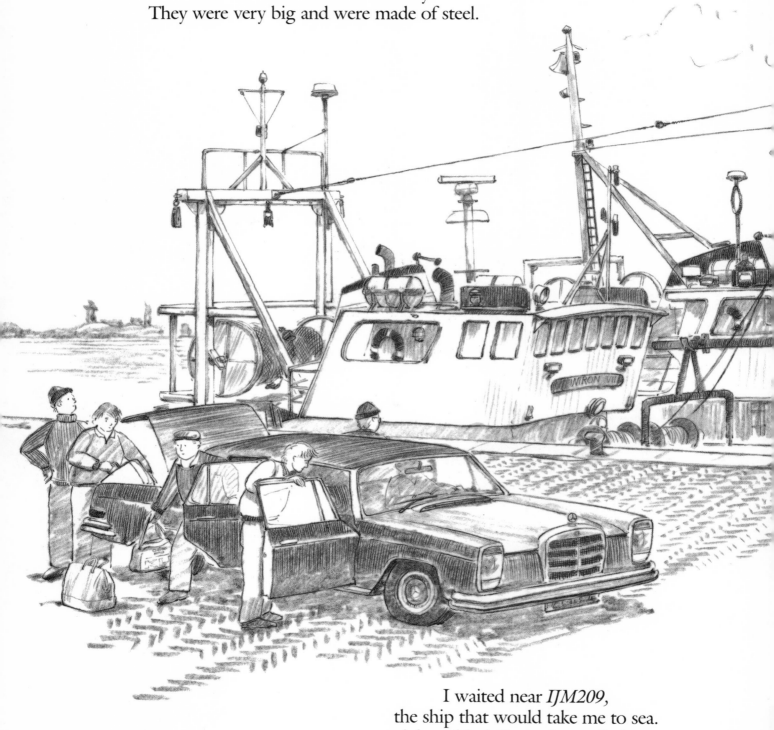

I waited near *IJM209*,
the ship that would take me to sea.
A long black limousine pulled up,
and a half-dozen men stepped out.

"Could they be the crew?" I wondered.

Indeed they were.
The skipper, named Wim, introduced himself
and showed me aboard.
"We won't go to sea today," he said,
pointing to the trawl.
"We ran over a wreck, so
we're going to do a little sewing!"

Wim's ship, IJM 209, is a typical Dutch cutter. It does not belong to Wim but to a fishing company that owns five different ships.

Ships on the North Sea are distinctive with their tall armored and roofed bow. This, as I would learn, is to safely break the strong waves.

So while the men knitted away on shore, I had time to study the ship.

Shuttle

It was nightfall before *IJM 209* cut the first waves with her tall armored bow.

IJM 209 *is 33 meters long by 6.74 meters wide (108.9 by 22.2 feet) and powered by an 800-horsepower engine.*

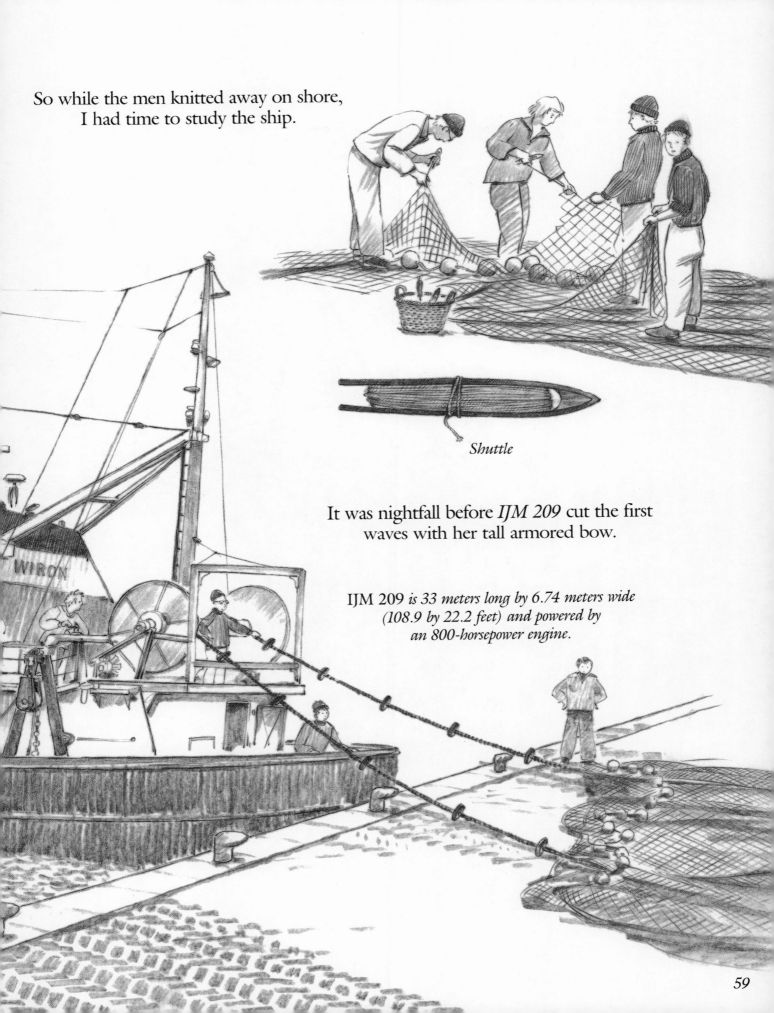

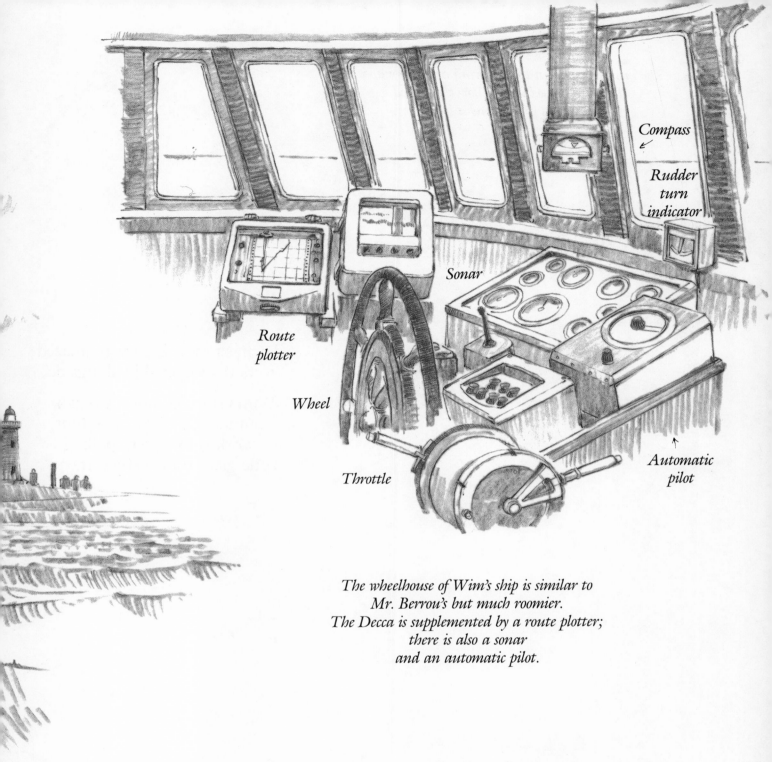

*The wheelhouse of Wim's ship is similar to
Mr. Berrou's but much roomier.
The Decca is supplemented by a route plotter;
there is also a sonar
and an automatic pilot.*

Wim told me that we would travel all night to get to the
fishing grounds, eight hours away, just off the coast of England.
We should be gone five days in all, if the weather was good.

I bid him good night and settled into my bunk,
wondering what the morning would bring.

The measurements in this drawing are approximate but give a good idea of how the paired trawling arrangement looks.

< 400 meters (1320 feet) >

< 50 meters (165 feet) >

< 700 meters (2310 feet) >

The next morning I was amazed to find another ship alongside.

Wim's ship did not trawl alone but in a pair. The two ships worked like oxen, pulling one giant trawl behind them.

I saw from Wim's map that the North Sea is virtually littered with wrecks. Little wonder after the numerous sea battles fought there ...and its notorious storms.

The wrecks posed quite a hazard for the trawl, and I imagined what it must look like under the waves.

30 meters (99 feet)

Although the crew of IJM 209 are all employees, they do not earn a fixed salary. Everyone is paid a percentage based on the earnings of the catch, like the crews in Brittany.

Two days went by in a rhythm similar to that of Mr. Berrou's ship in Brittany. The real difference was simply that everything was bigger.

The ship was bigger, the trawl was bigger, the catch was bigger, and, as I soon learned, the waves were bigger.

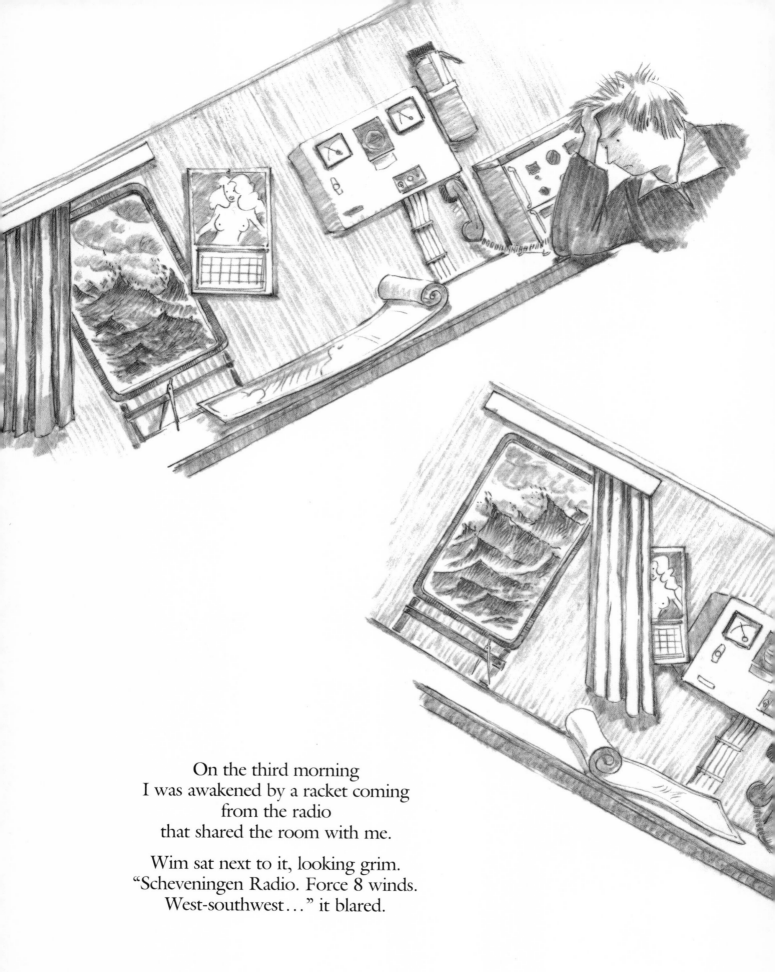

On the third morning
I was awakened by a racket coming
from the radio
that shared the room with me.

Wim sat next to it, looking grim.
"Scheveningen Radio. Force 8 winds.
West-southwest…" it blared.

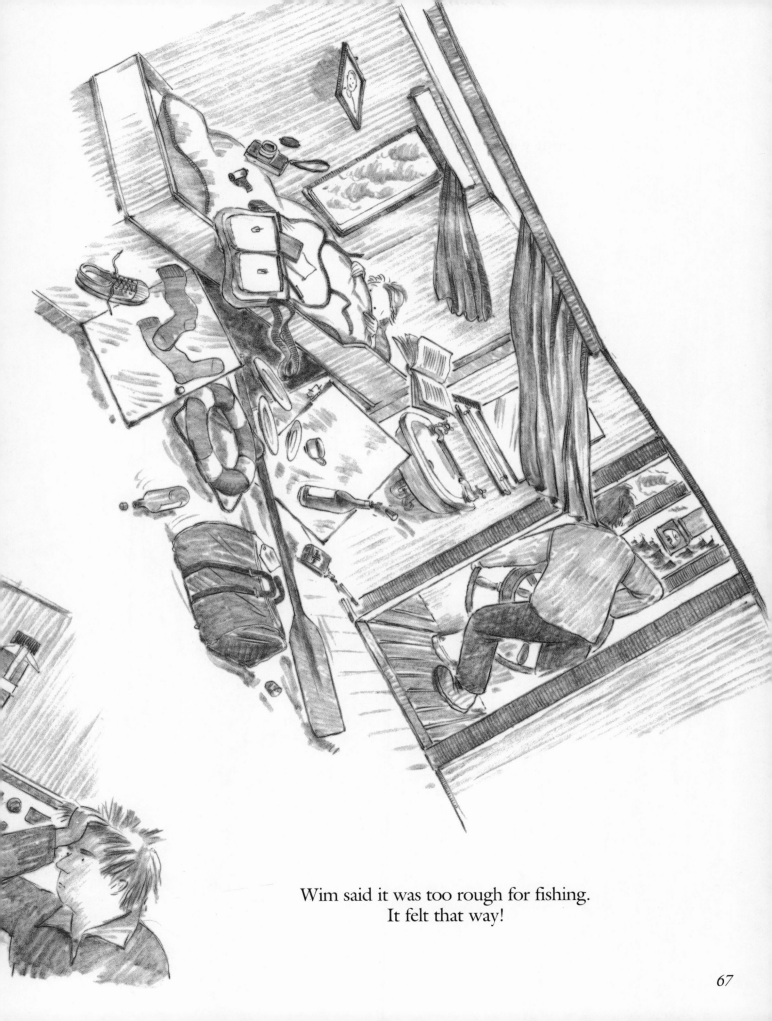

Wim said it was too rough for fishing.
It felt that way!

On day four fishing began again. But, alas for Wim, not for long. There was a strong tug from the trawl, and it was immediately obvious that we had just discovered an unknown wreck. There was nothing left to do but put back to port a day early.

It was fun to go to sea! But I must admit there are few feelings sweeter than stepping ashore again.

My sources:

Le Musée de la Pêche, Concarneau, Brittany.

Les Bateaux de Pêche de Bretagne, by Hervé Gloux and Jean-Yves Manac'h, Librairie Artème Fayard, 1976.

Fishing Gazette, New York, NY.

The Observer's Book of Ships, by Frank E. Dodman, Frederick Warne & Co. Ltd., 1970.

Les Pêches Maritimes (Que Sais-Je?), by Albert Boyer, Presses Universitaires de France, 1967.

Pêches et Pêcheurs de la Bretagne Atlantique, by Charles Robert-Muller, Librairie Armand Colin, Paris, 1944.

My thanks to:

Mr. Germain Berrou, for a trip aboard *La Madone,* Le Croisic;
Mssrs. Dany and Serge Debec, for a trip aboard *La Fleur des Vagues,* Le Croisic.

Mr. Guy Jacob, for a trip aboard *Kanedeven,* Le Quiberon;
Mr. Wim Ouwehand, for a trip aboard *IJM 209,* IJmuiden.

...and to

Mrs. Suzanne Gaudron,
Mr. Alain Lebreton,
Mr. Jean-Marie Le Faou,
Mr. Paul Brinkman,
Mr. Bernard Hilliet, Conserverie Belle-Iloise

...who all helped me aboard ship!

Huck Scarry, the son of children's book author/illustrator Richard Scarry, was born in Westport, Connecticut, where he spent his childhood. Later he moved to Switzerland with his parents. He studied art in Lausanne, Switzerland, and now lives in the old town of Geneva with his wife Marlis and their two young children. He has written and illustrated several other books for children, including *Life on a Barge: A Sketchbook* and *Balloon Trip: A Sketchbook,* both for Prentice-Hall.